WHY SOME COMPANIES BUILD MOMENTUM...
AND OTHERS DON'T

TURNING THE FLYWHEEL

A Monograph to Accompany
Good to Great

JIM COLLINS

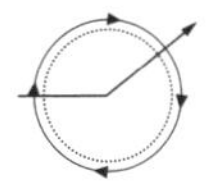

HarperCollins books may be purchased for educational, business, or sales promotional use. For information, please email the Special Markets Department at SPsales@harpercollins.com.

FIRST EDITION

Designed by Elements Design Group
Photograph of Jim Collins by George Lange

Library of Congress Cataloging-in-Publication Data has been applied for.

ISBN 978-0-06-293379-9

19 20 21 22 23 IMP 10 9 8 7 6 5 4 3 2 1

To my personal band of brothers — you know who you are —

in the spirit of loyalty, love, and enduring friendship.

TURNING THE FLYWHEEL

"Beauty does not come from decorative effects but from structural coherence." — Pier Luigi Nervi[1]

In the autumn of 2001, just as *Good to Great* first hit the market, Amazon.com invited me to engage in a spirited dialogue with founder Jeff Bezos and a few members of his executive team. This was right in the middle of the dot-com bust, when some wondered how (or if) Amazon could recover and prevail as a great company. I taught them about "the flywheel effect" that we'd uncovered in our research. In creating a good-to-great transformation, there's no single defining action, no grand program, no single killer innovation, no solitary lucky break, no miracle moment. Rather, it feels like turning a giant, heavy flywheel. Pushing with great effort, you get the flywheel to inch forward. You keep pushing, and with persistent effort, you get the flywheel to complete one entire turn. You don't stop. You keep pushing. The flywheel moves a bit faster. Two turns . . . then four . . . then eight . . . the flywheel builds momentum . . . sixteen . . . thirty-two . . . moving faster . . . a thousand . . . ten thousand . . . a hundred thousand. Then at some point—breakthrough! The flywheel flies forward with almost unstoppable momentum.

Once you fully grasp how to create flywheel momentum *in your particular circumstance* (which is the topic of this monograph) and apply that understanding with creativity and discipline, you get the power of strategic compounding. Each turn builds upon previous work as you make a series of good decisions, supremely well executed, that compound one upon another. This is how you build greatness.

The Amazon team grabbed onto the flywheel concept and deployed it to articulate the momentum machine that drove the enterprise at its best. From its inception, Bezos had infused Amazon with an obsession to create ever more value for ever more customers. It's a powerful animating force—perhaps even a noble purpose—but the key differentiator lay not just in "good intent" but in the way Bezos and company turned it into a repeating loop. As Brad Stone later wrote in *The Everything Store*, "Bezos and his lieutenants sketched their own virtuous cycle, which they believed powered their business. It went something like this: Lower prices led to more customer visits. More customers increased the volume of sales and attracted more commission-paying third-party sellers to the site. That allowed Amazon to get more out of fixed costs like the fulfillment centers and the servers needed to run the website. This greater efficiency then enabled it to lower prices further. Feed any part of this flywheel, they reasoned, and it should accelerate the loop." And so, the flywheel would turn, building momentum. Push the flywheel; accelerate momentum. Then repeat. Bezos, Stone continued, considered Amazon's application of the flywheel concept "the secret sauce."[2]

I've sketched my own take on the essence of the original Amazon flywheel in the nearby diagram. (Note: Throughout this monograph, I've included sketches of specific flywheels to illustrate the concept. To be clear, these reflect my own take on the flywheel from each case; the leaders who built these flywheels would likely draw them with more nuance than I have. Use these illustrative sketches to grasp the flywheel concept and to stimulate thinking about your own flywheel.)

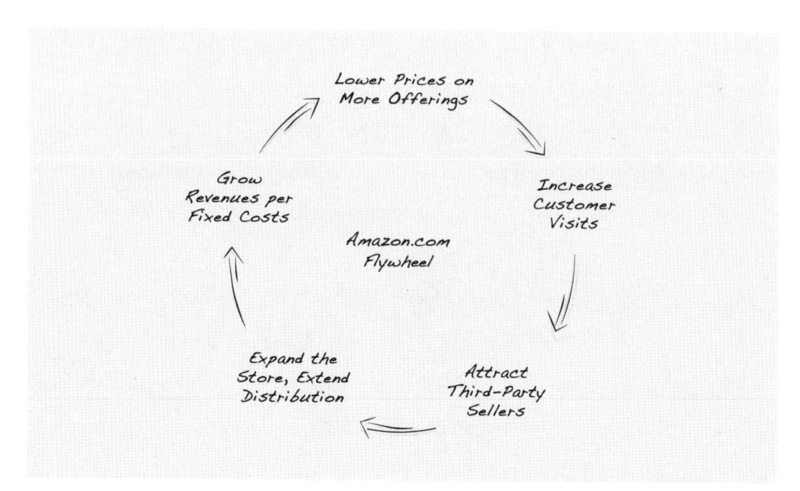

Notice the inexorable logic. Trace your way around the Amazon flywheel a few times in your mind, and you can almost get swept up in the momentum. Each component in the flywheel sets you up for the next component, indeed, almost throwing you around the loop.

Bezos and his team could have panicked during the dot-com bust, abandoned the flywheel, and succumbed to what I described in *Good to Great* as the doom loop. When caught in the doom loop, companies react to disappointing results without discipline—grasping for a new savior, program, fad, event, or direction—only to experience more disappointment. Then they react without discipline yet again, leading to even more disappointment. Instead, Amazon committed fully to its flywheel and then innovated aggressively within that flywheel to build and accelerate momentum. Amazon not only survived but also became one of the most successful and enduring companies to emerge from the dot-com era. Over time, Amazon would renew and extend the flywheel far beyond a simple e-commerce website and enhance the flywheel with new technology accelerators such as artificial intelligence and machine learning. But throughout, the underlying flywheel architecture remained largely intact, creating a customer-value compounding machine that many of the largest companies in the world came to fear.

> Never underestimate the power of a great flywheel, especially when it builds compounding momentum over a very long time. Once you get your flywheel right, you want to renew and extend that flywheel for years to decades—decision upon decision, action upon action, turn by turn—each loop adding to the cumulative effect. But to best accomplish this, you need to understand how *your specific flywheel* turns. Your flywheel will almost certainly not be identical to Amazon's, but it should be just as clear and its logic equally sound.

In the years since publishing *Good to Great*, I've challenged dozens of leadership teams to do for themselves what the Amazon team did for itself. Some of those teams traveled to our management lab at The Good to Great Project in Boulder, Colorado, and I watched each team assemble its flywheel, almost like putting together a jigsaw puzzle. They'd get the pieces laid out and then fiddle with them, arguing and debating, engaged in a disciplined thought process to get their flywheel right. What are the essential components? Which component comes first? What follows? Why? How do we complete the loop? Do we have too many components? Is anything missing? What evidence do we have that this works in practice? Gradually, their specific flywheel would start to emerge. When it all clicked, it felt like the final pieces of the jigsaw puzzle had popped into place. In clarifying their flywheels, these teams experienced the sense of excitement that comes when you see—and *feel*—how to generate the results necessary to achieve or extend a good-to-great breakthrough.

Bill McNabb, then CEO of the mutual fund giant Vanguard, brought his senior team to Boulder in 2009, and they worked for two days to crystallize their flywheel. They did an impressive job of capturing the essence of the Vanguard momentum machine, which I've sketched in a simplified flywheel diagram below.

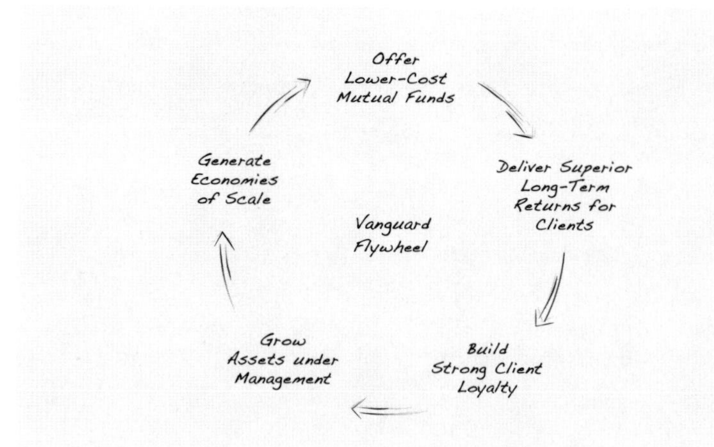

Notice how each component in the Vanguard flywheel isn't merely a "next action step on a list" but almost an *inevitable consequence of the step that came before.* If you offer lower-cost mutual funds, *you almost can't help* but deliver superior long-term returns to investors (relative to higher-cost funds that invest in the same assets). And if you deliver superior returns to investors, *you almost can't help* but build client loyalty. And if you build strong client loyalty, *you almost can't help* but grow assets under management. And if you grow assets under management, *you almost can't help* but generate economies of scale. And if you increase economies of scale, *you almost can't help* but have lower costs that you can pass along to clients. Vanguard had been turning some form of this flywheel for decades, built upon the insights and principles of its visionary founder Jack Bogle, who championed the world's first index mutual funds. But pausing to crystallize the underlying flywheel architecture gave the leadership team the clarity it needed to keep building momentum with fanatic intensity, especially coming out of the 2008–2009 financial crisis. From 2009 to 2017, Vanguard's flywheel continued to build momentum, more than doubling its assets under management to exceed $4 trillion.[3]

The Vanguard case exemplifies a key aspect of how the best flywheels work. If you nail one component, you're propelled into the next component, and the next, and the next, and the next—almost like a chain reaction. In thinking about your own flywheel, it's absolutely vital that it *not* be conceived as merely a list of static objectives that you've simply drawn as a circle. It must capture the *sequence* that ignites and accelerates momentum.

The intellectual discipline required to get the sequence right can produce profound strategic insight. As Stanford Graduate School of Business strategy professor Robert Burgelman once observed to a classroom full of students in 1982 (of which I was one), the greatest danger in business and life lies not in outright failure but in achieving success without understanding *why* you were successful in the first place. Burgelman's insight kept pinging in my brain throughout my twenty-five years of research into the question of what makes great companies tick, especially in explaining why some companies fall from grace. When you deeply understand the underlying causal factors that give your flywheel its momentum, you can avoid Burgelman's trap.

THE DURABILITY OF A GREAT FLYWHEEL

One of the biggest, and most common, strategic mistakes lies in failing to aggressively and persistently make the most of victories. One reason why some leaders make this mistake is that they become seduced by an endless search for the Next Big Thing. And sometimes they do find the Next Big Thing. Yet our research across multiple studies shows that if you conceive of your flywheel in the right way—and if you continually renew and extend the flywheel—it can be remarkably durable, perhaps even capable of carrying your organization through a major strategic inflection point or turbulent disruption. But to do so requires understanding the *underlying architecture of the flywheel as distinct from a single line of business or arena of activity.*

Let me use a classic historical case to illustrate, Intel's "dramatic" shift from memory chips to microprocessors. From its earliest days, Intel built a flywheel harnessing Moore's Law (the empirical observation that the number of components on an integrated circuit achieved at an affordable cost doubles roughly every eighteen months). From this insight, Intel's founding team created a strategic compounding machine: Design new chips that customers crave; price high before competition catches

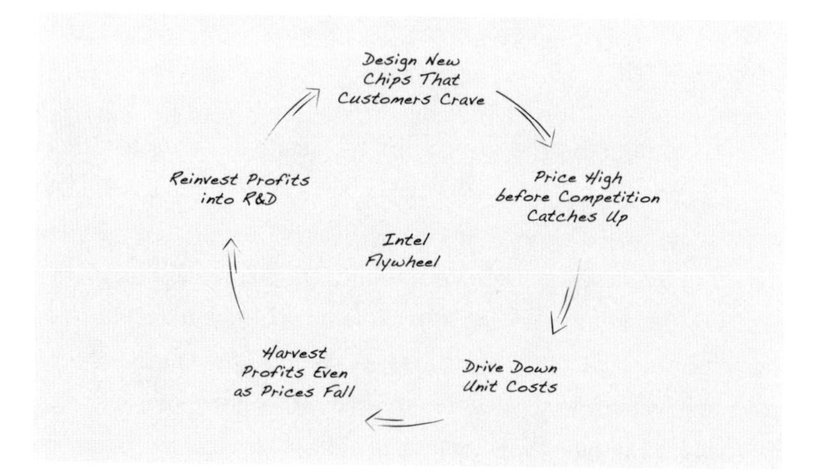

up; drive down unit costs as volume increases (due to economies of scale); harvest high profits even as competition drives down prices; and reinvest those profits into R&D to design the next generation of chips.

This flywheel powered Intel's rise from start-up to great company in the memory-chip business.[4]

Then in the mid-1980s, the memory-chip business careened into a brutal international price war. Intel's sales declined and profits evaporated. CEO Gordon Moore and President Andy Grove faced a stark reality: Intel's memory-chip business had become economically untenable and would remain so. In his must-read book *Only the Paranoid Survive*, Grove described an epiphany when he asked Moore, "If we got kicked out and the board brought in a new CEO, what do you think he would do?" Moore gave an unequivocal answer, "He would get us out of memories." So, Grove mused, "Why shouldn't you and I walk out the door, come back and do it ourselves?"[5] I carry an image in my mind of Grove and Moore pointing at each other and saying, "You're fired." Then walking out in the hall, pointing at each other and saying, "You're hired." Then walking back into the office as "new" leadership and saying, "That's it, we're getting out of memories!"

Now, consider the following question. In making this bold move, did Intel jettison its flywheel? *No!* Intel had been building up a side business in microprocessor chips for more than a decade, and the underlying flywheel architecture could apply just as soundly to microprocessor chips as memory chips. Different chips, to be sure, but very much the same underlying flywheel.

In 2002, I had a conversation with Grove about this very question in preparation for an on-stage conversation we were going to do together on the topic of building great companies. As we got to talking about the decision to get out of memories, Grove commented that, through the lens of the flywheel construct, Intel's bold memories-to-microprocessors shift wasn't quite as discontinuous as it appeared on the surface. It was really more of a transfer of momentum from memories to microprocessors, not a jagged break to create an entirely new flywheel. If Intel had tossed out its underlying flywheel architecture when it exited memories, it wouldn't have become the dominant chip maker that powered the personal computer revolution.

> For a truly great company, the Big Thing is never any specific line of business or product or idea or invention. *The Big Thing is your underlying flywheel architecture*, properly conceived. If you get your flywheel right, it can guide and drive momentum (with renewal and extensions) for at least a decade, and likely much longer. Amazon, Vanguard, and Intel didn't destroy their flywheels in response to a turbulent world; they disrupted the world around them by turning their flywheels.

This doesn't mean mindlessly repeating what you've done before. It means evolving, expanding, extending. It doesn't mean just offering Jack Bogle's revolutionary S&P 500 index fund; it means creating a plethora of low-cost funds in a wide range of asset categories that fit within the Vanguard flywheel. It doesn't mean just selling books online; it means expanding and evolving the Amazon flywheel into the biggest, most comprehensive e-commerce-store system in the world, and later extending that flywheel into selling its own devices (such as the Kindle and Alexa) and moving into physical retail (Amazon bought Whole Foods in 2017). It doesn't mean sticking doggedly with memory chips; it means redeploying the Intel flywheel to entirely new chip categories.

To be clear, my point is not that a flywheel will necessarily last forever. But just look at these three cases—Amazon, Vanguard, and Intel—each operating in a highly turbulent industry. In each company, the underlying flywheel propelled growth for decades. Intel did eventually evolve substantially beyond the chip business, but that doesn't change the fact that its initial flywheel architecture powered Intel's rise to a great company for more than thirty years. The logic underlying Vanguard's flywheel architecture remained essentially intact even as it approached the half-century mark. And at the time of this writing in 2018, the original Amazon flywheel has remained robust and relevant—thanks to renewal and extension—nearly two decades after it was first articulated.

Later in this text, I'll address how great companies renew and extend their flywheels. If you wake up one day to realize that your underlying flywheel no longer works, or that it's going to be disrupted into oblivion, then accept the fact that it must be recreated or replaced. But before you decide to toss out your flywheel, first make sure you understand its underlying architecture. Don't abandon a great flywheel when it would be a superior strategy to sustain, renew, and extend.

STEPS TO CAPTURING YOUR FLYWHEEL

So, then, how might you go about capturing your own flywheel? At our management lab, we've developed a basic process, refined during Socratic-dialogue sessions with a wide range of organizations. Here are the essential steps:

1. Create a list of significant replicable successes your enterprise has achieved. This should include new initiatives and offerings that have far exceeded expectations.
2. Compile a list of failures and disappointments. This should include new initiatives and offerings by your enterprise that have failed outright or fell far below expectations.
3. Compare the successes to the disappointments and ask, "What do these successes and disappointments tell us about the possible components of our flywheel?"
4. Using the components you've identified (keeping it to four to six), sketch the flywheel. Where does the flywheel start—what's the top of the loop? What follows next? And next after that? You should be able to explain *why* each component follows from the prior component. Outline the path back to the top of the loop. You should be able to explain how this loop cycles back upon itself to accelerate momentum.
5. If you have more than six components, you're making it too complicated; consolidate and simplify to capture the essence of the flywheel.
6. Test the flywheel against your list of successes and disappointments. Does your empirical experience validate it? Tweak the diagram until you can explain your biggest replicable successes as outcomes arising directly from the flywheel, and your biggest disappointments as failures to execute or adhere to the flywheel.

7. Test the flywheel against the three circles of your **Hedgehog Concept**. A Hedgehog Concept is a simple, crystalline concept that flows from deeply understanding the intersection of the following three circles: (1) what you're deeply passionate about, (2) what you can be the best in the world at, and (3) what drives your economic or resource engine. Does the flywheel fit with what you're deeply passionate about—especially the guiding core purpose and enduring core values of the enterprise? Does the flywheel build upon what you can be the best in the world at? Does the flywheel help fuel your economic or resource engine? (In the appendix to this monograph, I've created a short summary of the framework of concepts that have come from our research—concepts such as the Hedgehog Concept—along with a brief definition of each concept. This appendix also shows where the flywheel fits in the overall conceptual map for the journey from good to great. The first time I mention any of these concepts in the main text, I will put them in bold.)

Organizations without the components of a flywheel already in place—such as early-stage entrepreneurial companies—can sometimes jump-start the process by importing insights from flywheels that others have built. Jim Gentes founded Giro Sport Design on a new bicycle-helmet design that would be lighter and more aerodynamic than other helmets. Wearing a Giro helmet, the cyclist could ride faster, cooler, and safer. It would also be stylish and colorful, whereas other boxy helmets made the rider look like a geek monster from outer space in a B-grade 1950s horror film. After carrying a prototype to the Long Beach bike show, Gentes returned to his one-bedroom apartment with $80,000 of orders and began manufacturing batches of helmets in his garage.[6]

But how to turn a single product into a sustained flywheel, especially as a garage start-up? Gentes studied Nike and gleaned an essential insight. There's a hierarchy of social influence for athletic gear. If, for instance, you get a Tour de France winner to wear your helmet, serious nonprofessional cyclists will want to wear that helmet, which then starts the cascade of influence and builds the brand. Gentes validated this insight when he bet a substantial portion of the company's meager resources on sponsoring elite American cyclist Greg LeMond to wear a Giro helmet. In the dramatic finale of the 1989 Tour de France, everything came down to the final stage, a time trial into Paris. LeMond overcame a 50-second deficit at the start of the time trial to win the entire Tour by a mere 8 seconds—after a 23-day race—wearing an aerodynamic Giro helmet as he rocketed down the Champs-Élysées. Suddenly, it became very cool for serious riders to wear a helmet, so long as it was a Giro.[7]

And so, by adopting a key insight from Nike's flywheel and blending it with his own passion for inventing great new products, Gentes created a flywheel that propelled Giro far beyond the garage: Invent great products; get elite athletes to use them; inspire Weekend Warriors to mimic their heroes; attract mainstream customers; and build brand power as more and more athletes use the products. But then, to maintain the "cool" factor, set high prices and channel profits back into creating the next generation of great products that elite athletes want to use.

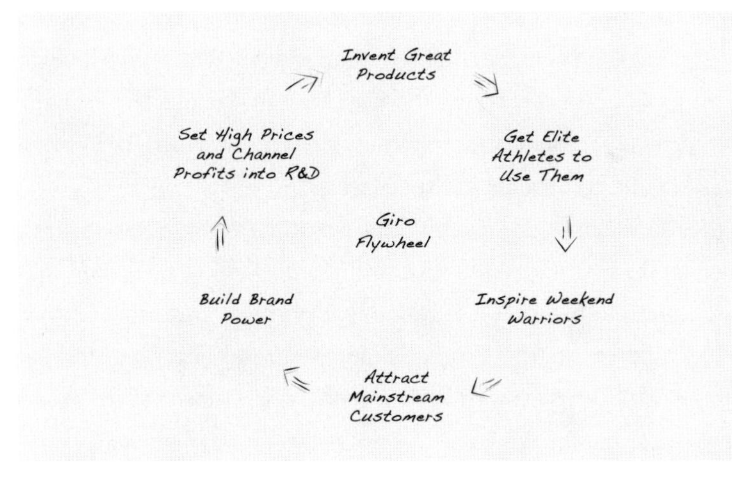

A flywheel need not be entirely unique. Two successful organizations can have similar flywheels. What matters most is how well you *understand* your flywheel and how well you *execute* on each component over a long series of iterations.

As Gerard Tellis and Peter Golder demonstrated in their book, *Will and Vision*, the pioneering innovators in a new business arena *almost never* (less than 10 percent of the time) become the big winners in the end. Similarly, across all our rigorous matched-pair research studies (*Built to Last, Good to Great, How the Mighty Fall,* and *Great by Choice*), we found no systematic correlation between achieving the highest levels of performance and being first into the game. This proved true even in innovation-intensive industries such as computers, software, semiconductors, and medical devices. Amazon and Intel started life in the wake of pioneers that preceded them; Advanced Memory Systems beat Intel to market in the early days of the DRAM-chip business, and Books.com preceded Amazon in online bookselling.[8] To be clear, the big winners in corporate history consistently surpassed a threshold level of innovation required to compete in their industries. But what truly set the big winners apart was their ability to turn initial success into a sustained flywheel, even if they started out behind the pioneers.[9]

NOT JUST FOR CEOS

Now, you might be thinking, "But I'm running a unit deep within a much larger organization. Can I build a flywheel?" Yes. To illustrate, let's look at a unit leader—an elementary school principal—who harnessed the flywheel effect within the walls of her individual school.

When Deb Gustafson became principal at Ware Elementary School, located on the Fort Riley Army base, she inherited one of the first Kansas public schools to be put "on improvement" for poor performance, with just one-third of students hitting grade level in reading. Not only did Gustafson struggle with a high student-mobility rate (due to transfers and deployments), but she also faced a 35-percent teacher-mobility rate.[10] And the children faced a special type of adversity, the stressful life of military families in wartime. It's one thing if your mom or dad has to travel for work; it's entirely another to see your mom or dad deployed to a combat zone. *These kids don't have time to wait*, Gustafson told herself. If we fail them at first grade or second grade, if they leave our school unable to read, we've failed them for the rest of their lives. *We simply cannot fail.*

Teaching is a relationship, not a transaction, and Gustafson believed that relationships could be built only on a foundation of collaboration and mutual respect. When parents are being shipped off to war, when families must sacrifice in service to country, the last thing kids need is warring factions inside their school. They need to feel a sense of calm, that the staff is there for them and is united in a mission to support them. Gustafson later described how she immediately grasped the applicability of the flywheel concept to her school when she read *Good to Great and the Social Sectors.* "When I got to the part about turning the flywheel, I was bouncing up and down out of my seat," said Gustafson. "I love the idea that if you can get everyone pushing the flywheel, all going in the same direction, it just starts working automatically."

Gustafson didn't wait for the district superintendent or the Kansas Commissioner of Education or the U.S. Secretary of Education to fix the entire K–12 system's flywheel. She threw herself into creating a unit-level flywheel right there in her individual school.

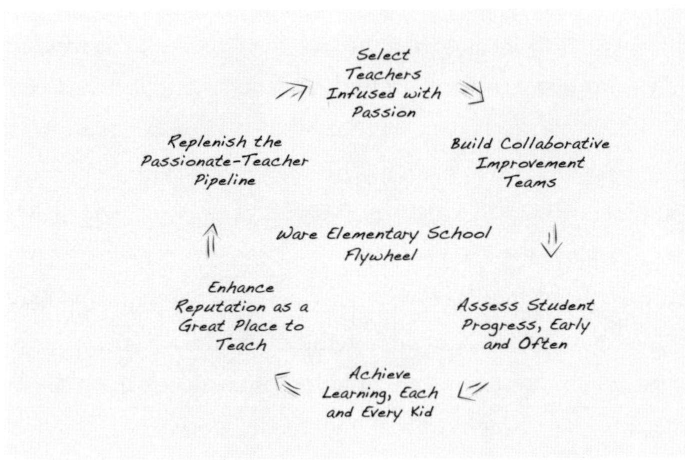

Flywheel step 1: Select teachers infused with passion. "We could not easily attract experienced teachers to teaching on a military base in rural Kansas," explained Gustafson. "So, I focused on passionate potential, even if inexperienced, figuring that people with the right values and irrepressible enthusiasm could be harnessed and shaped into effective teachers." All that passionate energy pulsating through the halls got the flywheel going, but it had to be guided, channeled, harnessed; it would simply make no sense to just throw inexperienced teachers into the classroom completely unprepared. That drove Gustafson to flywheel step 2: Build collaborative improvement teams. Each teacher would join a team led by an experienced Ware teacher who exemplified the culture. The mechanism generated cohesion and momentum as teams met in collaborative improvement meetings at least once per week, teachers coming together to share ideas, get feedback, discuss individual student progress, and improve the Ware teaching recipes.

But, of course, you can improve only if you know how you're doing and how each child is progressing. And that threw Ware right into flywheel step 3: Assess student progress, early and often. A continuous stream of data, shared and discussed in teams, generated energy—*We have to succeed for every child! We can't let any child fall behind! Each kid matters!* Teachers and teams set goals and crafted specific plans to help children who might be falling behind. The momentum increased as teams met quarterly with school leaders to further refine student plans and keep the flywheel spinning toward step 4: Achieve learning, each and every kid. Gustafson and her teachers took a school in which fewer than 35 percent of students were reading at a satisfactory level and changed the trajectory: They hit 55 percent at the end of year 1, 69 percent at year 3, 96 percent at year 5, then 99 percent at years 7, 8, 9, and beyond.[11]

All this fed right into flywheel step 5: Enhance the school's reputation, not just for results but also as a great place to teach. And that, in turn, brought the flywheel around to step 6: Replenish the passionate-teacher pipeline. Along the way, Ware earned status as a professional-development school at Kansas State University, further feeding the flywheel with a continuous stream of student teachers and interns. "We'd get passionate people with teaching potential into the building, and they'd fall in love with our school," explained Gustafson. "It's about the culture, and the relationships, and the collaboration with your teammates to improve and deliver for the kids—all that made us attractive to the right people. And that kept the pipeline of passionate people flowing so that we could turn the flywheel year after year after year." At the time of this writing, the Ware flywheel Gustafson created had been turning for more than fifteen years, touching as many as nine hundred military children per year.[12]

Leaders who create pockets of greatness at the unit level of their organization—leaders like school principal Gustafson—don't sit around hoping for perfection from the organization or system around them. They figure out how to harness the flywheel effect within their unit of responsibility. No matter what your walk of life, no matter how big or small your enterprise, no matter whether it's for-profit or nonprofit, no matter whether you're CEO or a unit leader, the question stands, *How does your flywheel turn?*

You'll find the flywheel effect in social movements and sports dynasties. You'll find the flywheel effect in monster rock bands and the greatest movie directors. You'll find the flywheel effect in winning election campaigns and victorious military campaigns. You'll find the flywheel effect in the most successful long-term investors and in the most impactful philanthropists. You'll find the flywheel effect in the most respected journalists and the most widely read authors. Look closely at any truly sustained great enterprise and you'll likely find a flywheel at work, though it might be hard to discern at first.

Before I move on to the question of how to think about changing and extending a flywheel, let me illustrate how far afield the flywheel principle can apply. I'll close out this section with a highly creative nonprofit, the Ojai Music Festival, which produces an annual musical adventure performed by some of the best musicians and composers from around the world in a magical place.

The Ojai flywheel cycle starts with attracting unconventional and exceptional talent. Each year, a different music director assumes responsibility as the chief musical curator. From composers like Igor Stravinsky and Pierre Boulez in its early days to violinist Patricia

Kopatchinskaja and pianist Vijay Iyer in the contemporary era, each music director brings his or her own distinctive genius, sparking creative renewal right from the get-go.[13] It's as if the festival puts up a blank canvas with an unstated challenge—all we ask is that you paint a masterpiece. Except that instead of a painting on a canvas, the masterpiece composition is a musical experience that engulfs artists and audience alike. "We've been able to attract unconventional talent to Ojai for two key reasons," explained Tom Morris, artistic director of the festival for nearly two decades. "First, they're energized by whom they get to play with, and second, they're energized by the fact that we unleash their creativity. It's like a big snow globe; you shake it up and see what comes down."[14]

The next flywheel step flows from a rigorous constraint. The festival lasts just four days, period. All that transcendent creativity, the snow globe of swirling ideas, must be forged into a tight program. Most ideas—even many great ideas—have to be cut in the end. And that brings us to the crucial insight, the causal link that snapped the flywheel around, from wild creativity to enhanced community support. "We don't want to evoke an appreciative audience response," Morris explained. "We want to *provoke* a passionate audience *reaction*."

Morris tells the story of a town resident who hadn't attended the festivals because he didn't like "that kind of music." But one day, the resident happened to walk into a festival performance of "Inuksuit," a spatial piece for nine to ninety-nine percussionists composed by John Luther Adams. By "walk into," I don't mean that he walked into the back of a concert hall with an orchestra far away on stage; he quite literally walked *into* the middle of the performance, with the players spread throughout a town park as they played amidst groves and paths, and the audience *surrounded* by sounds coming from all sides. There were tom-toms, cymbals, triangles, glockenspiels, sirens, piccolos, and all sorts of drums of various sizes and shapes. The music gradually rose from quiet to raucous, then began quieting down until it tapered off to its conclusion, seamlessly supplanted by the chirping of local birds chiming in from the trees. As players periodically moved to different stations throughout the park, some even climbing up into trees while the audience roamed and milled about, the unfolding performance enveloped them all.[15] Snap-click went the flywheel, and the once-skeptic who didn't like "that type of music" found himself transfixed by the experience and transformed into a passionate supporter of the festival.[16]

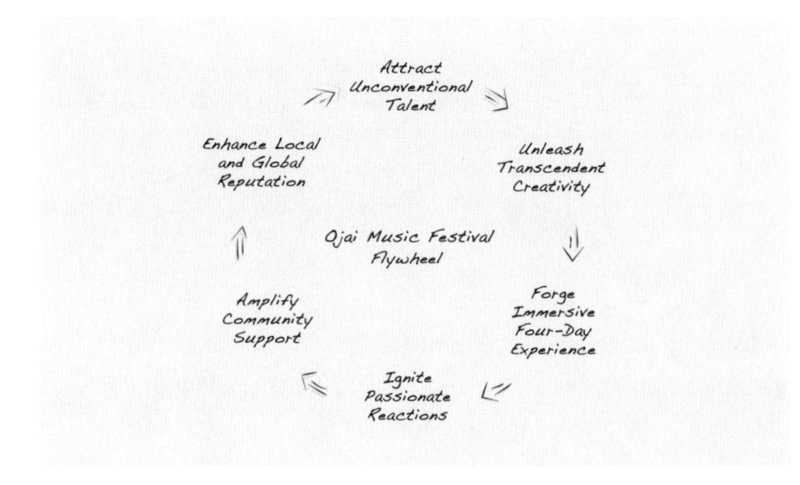

Morris and his colleagues understood that the most committed audience members want to be engaged, inspired, challenged, surprised, stunned, overwhelmed. They don't want to have a "nice listening moment" that they forget. They want to grow from a transformative musical experience that ignites the spirit and has a lasting emotional impact. And each time the festival delivered on that promise, the flywheel spun around, fueling the resource engine, building Ojai's reputation, and attracting the next wave of unconventional talent to create the next masterpiece and turn the flywheel anew.

EXECUTE AND INNOVATE—RENEWING THE FLYWHEEL

Once you get the flywheel right, the question becomes, what do we need to do better to accelerate momentum? The very nature of a flywheel—that it depends upon getting the sequence right and that every component depends on all the other components—means that *you simply cannot falter on any primary component and sustain momentum.* Think of it this way. Suppose you have, say, six components in the flywheel, and you score your performance in each from 1 to 10. What happens if your execution scores are 9, 10, 8, 3, 9, and 10? The entire flywheel stalls at the component scoring 3. To regain momentum, you need to bring that 3 up to at least an 8.

> The flywheel, when properly conceived and executed, creates *both* continuity *and* change. On the one hand, you need to stay with a flywheel long enough to get its full compounding effect. On the other hand, to keep the flywheel spinning, you need to continually renew, and improve each and every component.

In *Built to Last*, Jerry Porras and I observed that those who build enduring great companies reject the "Tyranny of the OR" (the view that things must be either A OR B but not both). Instead, they liberate themselves with the "**Genius of the AND**." Instead of choosing between A OR B, they figure out a way to have both A AND B. When it comes to the flywheel, you need to fully embrace the Genius of the AND, *sustain* the flywheel AND *renew* the flywheel.

The Cleveland Clinic became one of the most admired healthcare institutions in the world by embracing the Genius of the AND—consistency AND change—in its flywheel. The flywheel traces its roots back to the clinic's founding, when three physicians served in World War I and came away inspired by military teamwork. When you're serving soldiers carried off the battlefield, you don't ask, "Hey, what's my reimbursement rate? Am I going to get a bonus for this?" You work shoulder to shoulder with your colleagues, throwing whatever skills you have into the mix to save as many lives as possible and get them home to the people they love.

From this life-shaping experience, the three physicians vowed to create a distinctive new medical center after the war, one with a highly collaborative culture filled with people utterly obsessed with caring for patients. From its inception, the Cleveland Clinic focused on attracting first-rate physicians who would work on salary—no incentives based on the number of patients or procedures—because they'd be motivated primarily by working with world-class colleagues with a singular goal, *do what's best for the patient*. The Cleveland Clinic flywheel begins with the right people operating in a culture that drives patient outcomes, which then feeds into attracting patients and building the resource engine, which can then be redeployed to build capabilities and attract more of the right people to drive the flywheel around.[17]

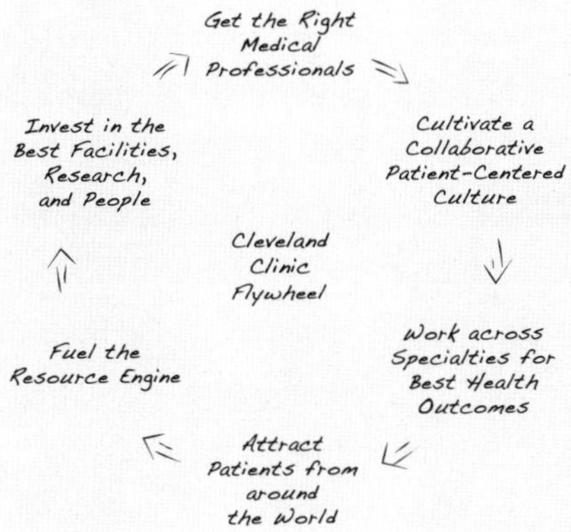

When Dr. Toby Cosgrove became CEO of the Cleveland Clinic in 2004, he deeply understood both the spirit and the logic of the flywheel. A military physician as a young man, he'd been deployed to the Vietnam War and was put in charge of a hospital; like the founders, he'd learned firsthand about working in teams and mobilizing all sorts of people with different skills to get things done in the chaos of incoming battlefield casualties. He joined the Cleveland Clinic in 1975 as a cardiac surgeon and led its heart program to a number-one ranking in *U.S. News & World Report*. Yet even with all this success, Cosgrove sensed that the Cleveland Clinic needed to rededicate itself to the proposition that the patient must come first. He challenged himself and his colleagues to address what needed to be changed, improved, and created to better serve the patient. For instance, they realized that a traditional structure organized by competencies (surgery, cardiology, etc.) favored medical tradition over working across specialties to best serve the patient. So, they instituted a structural change, creating institutes around patient needs, such as the Miller Family Heart & Vascular Institute that housed physicians from all the relevant specialties in the same location.

In his book, *The Cleveland Clinic Way*, Cosgrove details the myriad of changes put in place to renew the flywheel—changes big and small, strategic and tactical, structural and symbolic. From 2004 to 2016, the flywheel gained a huge burst of momentum—doubling revenues, patient visits, and research funding—while the Cleveland Clinic exported its brand across a growing network, from Ohio to Florida to Abu Dhabi. They renewed every component of the flywheel, *but they didn't dismantle it.* "Underneath, it's the original flywheel," said Cosgrove. "We reinvigorated it."[18]

> There are two possible explanations for a stalled or stuck flywheel. Possible explanation #1: The underlying flywheel is just fine, but you're failing to innovate and execute brilliantly on every single component; the flywheel needs to be reinvigorated. Possible explanation #2: The underlying flywheel no longer fits reality and must be changed in some significant way. It's imperative that you make the right diagnosis.

Over the long course of time (multiple decades), a flywheel might evolve significantly. You might replace components. You might delete components. You might revise components. You might narrow or broaden the scope of a component. You might adjust the sequence. These changes might happen by a process of invention, as you discover or create fundamentally new activities or businesses. Or they might happen by a process wherein you **confront the brutal facts** and practice **productive paranoia** about existential threats to your flywheel. For example, a company whose business model depended on collecting the personal information of millions of people found its flywheel imperiled by a data breach. Members of the executive team realized that they needed to insert a component dedicated to protecting privacy and earning trust. The rest of the flywheel remained intact, but without this vital new component, the company might have woken up one day on the verge of extinction.

That said, if you feel compelled to continuously make fundamental changes to the sequence or components of the flywheel, you've likely failed to get your flywheel right in the first place. Rarely does a great flywheel stall because it's run out of potential or is fundamentally broken. More often, momentum stalls due to either poor execution and/or failure to renew and extend within a fundamentally sound flywheel architecture. It is to the topic of extending the flywheel that we now turn.

EXTENDING THE FLYWHEEL

How do great companies go about extending a flywheel? The answer lies in a concept I developed with my colleague Morten Hansen in our book, *Great by Choice*. Morten and I systematically studied small entrepreneurial companies that became the 10X winners (beating their industries by more than ten times, in returns to investors) in highly turbulent industries in contrast to less successful comparison cases in the same environments. We found that both sets of companies made big bets but with a huge difference. The big successes tended to make big bets *after* they'd empirically validated that the bet would pay off, whereas the less successful comparisons tended to make big bets before having empirical validation. We coined the concept **fire bullets, then cannonballs** to capture the difference.[19]

Here's the idea: Imagine a hostile ship bearing down on you. You have a limited amount of gunpowder. You take all your gunpowder and use it to fire a big cannonball. The cannonball flies out and splashes in the ocean, missing the oncoming ship. You turn to your stockpile and discover that you're out of gunpowder. You're in trouble. But suppose instead that when you see the ship bearing down, you take a little bit of gunpowder and fire a bullet. It misses by 40 degrees. You make another bullet and fire. It misses by 30 degrees. You make a third bullet and fire, missing by only 10 degrees. The next bullet hits—ping!—the hull of the oncoming ship. You have empirical validation, a calibrated line of sight.

Now, you take all the remaining gunpowder and fire a big cannonball along the calibrated line of sight, which sinks the enemy ship.

> In looking across the history of great companies in all our research studies, we find a frequent pattern. They usually begin life being successful in a specific business arena, making the most of their early big bets. But soon they make a conceptual shift from "running a business" to *turning a flywheel.* And over time, they extend that flywheel by firing bullets, then cannonballs. They crank the flywheel in their first arena of success, while simultaneously firing bullets to discover new things that might work, and as a hedge against uncertainty.

Some bullets hit nothing, but some give enough empirical validation that the company then fires a cannonball, providing a big burst of momentum. In some cases, these extensions come to generate the vast majority of momentum in the flywheel, and in a few cases (such as when Intel moved from memories to microprocessors), they entirely replace what came before.

Apple's flywheel extension into its biggest business—smart handheld devices—followed exactly this pattern. In 2002, nearly all of Apple's flywheel momentum came from its line of Macintosh personal computers. But it had fired a bullet on this little thing called an iPod, described in its 2001 form 10-K as simply "an important and natural extension" of Apple's personal computer strategy. In 2002, the iPod generated less than 3 percent of Apple sales. Apple kept firing bullets on the iPod, developing an online music store along the way (iTunes). The bullets kept hitting, the iPod kept adding momentum to the flywheel, and Apple eventually fired a huge cannonball, betting big on the iPod and iTunes. And Apple kept extending the flywheel, from iPod to iPhone, from iPhone to iPad, and Apple's flywheel extension became the largest generator of momentum.[20]

In the nearby table, I've listed a range of fabulous flywheel extensions from corporate history. In every case, the company followed the bullet-

COMPANY	FIRST ARENA OF FLYWHEEL SUCCESS	NEXT BIG EXTENSION OF THE FLYWHEEL
3M	Abrasives (e.g., sandpaper)	Adhesives (e.g., Scotch Tape)
Amazon	Internet-enabled retail for consumers	Cloud-enabled web services for enterprises
Amgen	Therapeutics for low-blood-cell conditions	Therapeutics for inflammation and cancer
Apple	Personal computers	Smart handhelds (iPod, iPhone, iPad)
Boeing	Military aircraft	Commercial jetliners
IBM	Accounting tabulating machines	Computers
Intel	Memory chips	Microprocessors
Johnson & Johnson	Medical and surgical products	Consumer health-care products
Kroger	Small-scale grocery stores	Large-scale superstores
Marriott	Restaurants	Hotels
Merck	Chemicals	Pharmaceuticals
Microsoft	Computer languages	Operating systems and applications
Nordstrom	Shoe stores	Department stores
Nucor	Steel joists	Manufactured steel
Progressive	Non-standard (high-risk) car insurance	Standard car insurance
Southwest Airlines	Low-cost intrastate airline (Texas only)	Low-cost interstate airline (coast to coast)
Stryker	Hospital beds	Surgical products
Walt Disney	Animated films	Theme parks

to-cannonball method to extend and accelerate an underlying flywheel that had been turning for years.

When does a new activity become a second flywheel, as distinct from an extension? Most seeming "second flywheels" come about organically, as bullet-to-cannonball extensions of a primary flywheel. Amazon showed this precise pattern with its Amazon Web Services, which enables organizations big and small to efficiently buy computing power, store data, host websites, and avail themselves of other technology services. Amazon Web Services began as an internal system to provide backend technology support for Amazon's own e-commerce efforts. In 2006, the company fired a bullet on offering these very same services to outside companies. The bullet hit its target and Amazon had enough calibration to fire a cannonball. A decade later, Amazon Web Services (while still contributing less than 10 percent of Amazon's overall net sales) generated a substantial portion of Amazon's operating income.[21]

Even though Amazon Web Services first appears to be a very different activity than the consumer-retail business, it has substantial similarities. As Bezos wrote in his 2015 annual letter to shareholders, "Superficially, the two could hardly be more different. One serves consumers and the other serves enterprises . . . Under the surface, the two are not so different after all." Amazon Web Services aims to lower prices and expand offerings to an ever-growing cadre of customers, leading to increasing revenues per fixed costs, which then drives the flywheel around again. The whole idea is to make it as easy and cost-effective for enterprises to meet their technology needs as it is for consumers buying personal stuff at the Amazon marketplace. Sure, there are differences in how the two businesses operate, but the two are more like fraternal twins than being from an entirely different family lineage.

> Every large organization will eventually have multiple sub-flywheels spinning about, each with its own nuance. But to achieve greatest momentum, they should be held together by an underlying logic. And each sub-flywheel should clearly fit within and contribute to the whole.

The most important thing is to keep turning the overall flywheel—and every component and sub-flywheel—with creative intensity and relentless discipline. Even with the early growth and profitability of Amazon Web Services, Bezos remained obsessed with keeping Amazon's consumer-retail business as vibrant and energized as when the company first began. After all, even as Amazon approached $200 billion in annual revenues, it had less than 1 percent of the global retail market.[22]

STAY ON THE FLYWHEEL . . . AND OUT OF *HOW THE MIGHTY FALL*

In studying the horrifying fall of once-great companies, we see them abandoning the key principles that made them great in the first place. They vest the wrong leaders with power. They veer from the **First Who** principle and cease to **get the right people on the bus**. They fail to confront the brutal facts. They stray far beyond the three circles of their Hedgehog Concept, throwing themselves into activities at which they could never become best in the world. They subvert discipline with bureaucracy. They corrupt their core values and lose their purpose. And one of the biggest patterns exhibited by once-great companies that bring about their own senseless self-destruction is failure to adhere to the flywheel principle.

In our research for *How the Mighty Fall*, we found that the demise of once-great companies happens in five stages: (1) Hubris Born of Success, (2) Undisciplined Pursuit of More, (3) Denial of Risk and Peril, (4) Grasping for Salvation, and (5) Capitulation to Irrelevance or Death. Take special note of Stage 4, Grasping for Salvation. When companies fall into Stage 4, they succumb to the doom loop, the exact opposite of building flywheel momentum. They grasp for charismatic saviors or untested strategies or big uncalibrated cannonballs or cultural revolutions or "game-changing" acquisitions or transformative technologies or radical restructurings (then another and another) or... well, you get the idea.

> In Stage 4, each grasp for salvation creates a burst of hope and momentary momentum. But if there's no underlying flywheel, the momentum doesn't last. And with each grasp, the enterprise erodes capital—financial capital, cultural capital, stakeholder capital—and weakens. If the company never gets back to the discipline of the flywheel, it will likely continue to spiral downward until it enters Stage 5. No enterprise comes back from Stage 5. Game over.

Circuit City, which we studied in the original research for *Good to Great*, later "earned" a spot in *How the Mighty Fall*, and its demise teaches

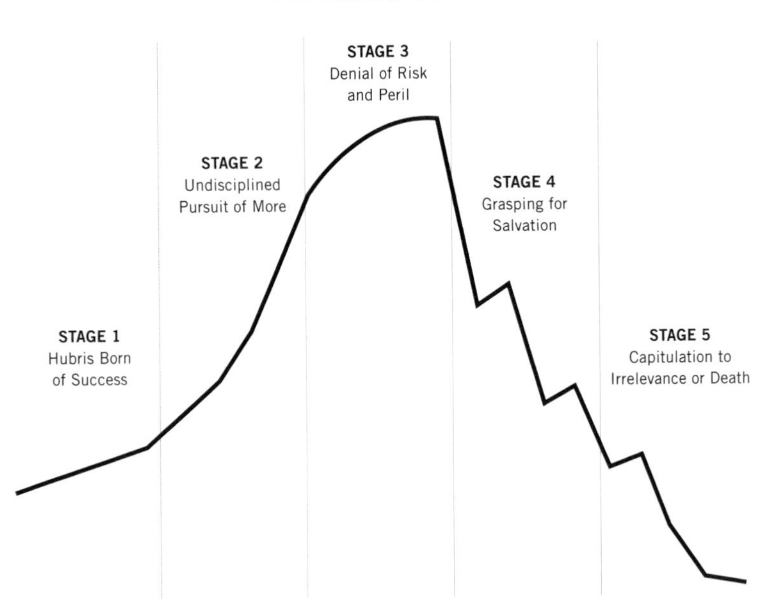

important lessons about the flywheel. During its good-to-great years, Circuit City made the leap from dismal mediocrity to superstar success under the inspired **Level 5 leadership** of Alan Wurtzel, transforming a hodgepodge of hi-fi stores into a sophisticated system of consumer electronics superstores, generating total returns to investors of more than eighteen times the general stock market over fifteen years. But after the Wurtzel era, Circuit City began to decline, slowly at first, almost imperceptibly, as usually happens with companies moving through the early stages of decline, then it plummeted precipitously through Stage 4, and right into Stage 5 capitulation and death.

How did this happen? A big part of the answer lies in two fundamental mistakes the post-Wurtzel leadership team made related to the flywheel principle. First, they became distracted by searching for the Next Big Thing. Circuit City sought big new ideas for growth, anticipating the day that the consumer electronics superstores would run out of great locations in which to open across the country. This in itself was a good idea, just as Amazon continually sought new ideas to propel the flywheel. But, unlike Amazon under Bezos, Circuit City neglected to keep the consumer electronics retail business robust and relevant. Meanwhile, an up-start competitor named Best Buy seized the market.[23]

Second—and this is the most fundamental lesson from Circuit City's demise—the post-Wurtzel team underestimated how far a flywheel could go if seen as an underlying architecture (that can be extended) rather than as a single line of business. The great tragedy of Circuit City is that it did indeed invent a spectacular extension, called CarMax, which could have created momentum for at least another couple of decades. The idea behind CarMax was to do for the used-car business what the Wurtzel team had done for hi-fi stores, to professionalize and transform a hodgepodge industry into a sophisticated system of superstores under one trusted brand.[24]

Circuit City fired a bullet with its first CarMax store in Richmond, Virginia. It proved successful. So, it fired a second bullet, opening a second CarMax in Raleigh, North Carolina, which also proved successful. Next, it fired two more bullets in Atlanta, Georgia. With empirical validation in hand, Circuit City fired a cannonball, opening CarMax superstores and expanding into new regions—Florida, Texas, California, and beyond. By the early 2000s, CarMax was growing at close to 25 percent a year, generating more than $3 billion in profitable sales in 2002.[25]

Now, stop and think about this for a minute. How did CarMax's success presage Circuit City's fall? With CarMax, Circuit City had created a huge new extension of its flywheel that could generate years of additional momentum. The CarMax flywheel extension could have been analogous to Apple's flywheel extension from personal computers to smart handhelds, Boeing's flywheel extension from military propeller-driven bombers to commercial jet airliners, Marriott's flywheel extension from restaurants to hotels, and Walt Disney's flywheel extension from animated films to theme parks. And in the event that the consumer electronics superstores did become untenable as a business, the company could redeploy all its energy into CarMax (similar to Intel's move out of memories and into microprocessors). But to do so would have required the conceptual wisdom to see CarMax as an extension of an underlying flywheel architecture.

Sadly, the post-Wurtzel team got rid of the CarMax superstores, jettisoning CarMax into its own separate company. It was as if Intel had decided in 1985 to get rid of the microprocessor business and keep the memory-chip business; the spun-out microprocessor company might have been successful, but Intel would have likely died. Fortunately for Intel, Grove and Moore had the strategic acumen to see the microprocessor business as an extension of its underlying flywheel. Circuit City failed to make this conceptual leap.

As Alan Wurtzel later wrote in his book, *Good to Great to Gone* (which I warmly recommend reading), "Looked at from a long-term perspective, it is disappointing that CarMax did not remain part of the Circuit City portfolio . . . The initial premise for CarMax was to create a portfolio of retail companies, so that as one matured another would be coming along to support the growth of the whole."[26] Wurtzel understood CarMax as part of a bigger flywheel, but those who ran the company after him didn't. If Circuit City had continued to evolve and renew the consumer electronics superstores (as Best Buy did) and continued to extend its underlying flywheel into new arenas (such as with CarMax), it might have remained a great company, climbing steadily upward in the S&P 500. Instead, Circuit City lost all its flywheel momentum and careened into the later stages of decline—down, down, down, the doom loop hurtling the company toward irrelevance. The once good-to-great company died in the winter of 2008.[27]

THE VERDICT OF HISTORY

After conducting a quarter-century of research into the question of what makes great companies tick—more than six thousand years of combined corporate history in the research data base—we can issue a clear verdict. The big winners are those who take a flywheel from ten turns to a billion turns rather than crank through ten turns, start over with a new flywheel, push it to ten turns, only to divert energy into yet another new flywheel, then another and another. When you reach a hundred turns on a flywheel, go for a thousand turns, then ten thousand, then a million, then ten million, and keep going until (and unless) you make a conscious decision to abandon that flywheel. Exit definitively or renew obsessively, but never—ever—neglect your flywheel. Apply your creativity and discipline to each and every turn with as much intensity as when you cranked out your first turns on the flywheel, nonstop, relentlessly, ever building momentum. If you do this, your organization will be much more likely to stay out of *How the Mighty Fall* and earn a place amongst those rare few that not only make the leap from good to great but also become built to last.

APPENDIX

THE FLYWHEEL WITHIN A FRAMEWORK

A Map for the Journey from Good to Great

I wrote this monograph to share practical insights about the flywheel principle that became clear in the years after first writing about the flywheel effect in chapter 8 of *Good to Great*. I decided to create this monograph because I've witnessed the power of the flywheel, when properly conceived and harnessed, in a wide range of organizations: in public corporations and private companies, in large multinationals and small family businesses, in military organizations and professional sports teams, in school systems and medical centers, in investment firms and philanthropic endeavors, in social movements and nonprofits.

Yet the flywheel effect alone does not make an organization great. The flywheel fits within a framework of principles we uncovered through more than a quarter-century of research into the question of what makes a great company tick. We derived these principles using a rigorous matched-pair research method, comparing companies that became great with companies (in similar circumstances) that did not. We'd systematically analyze the histories of the contrasting cases and ask, "What explains the difference?" (See nearby diagram, "The Good-to-Great Matched-Pair Research Method.")

THE GOOD-TO-GREAT MATCHED-PAIR RESEARCH METHOD

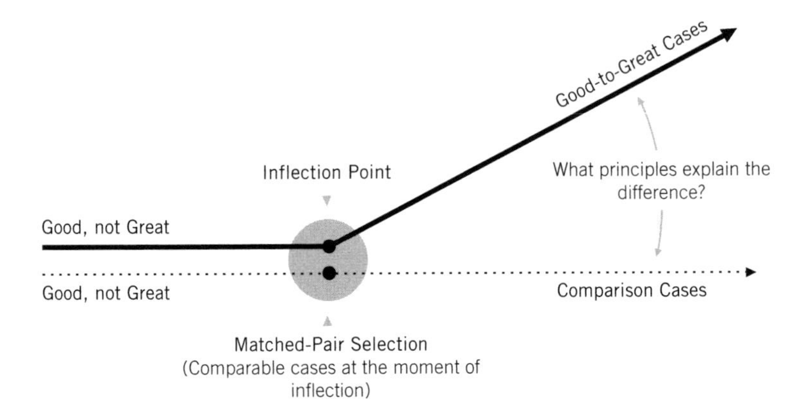

My research colleagues and I applied the historical matched-pair research method in four major studies, each using a different lens, that resulted in four books: *Built to Last* (coauthored with Jerry I. Porras), *Good to Great, How the Mighty Fall,* and *Great by Choice* (coauthored with Morten T. Hansen). We also extended the principles beyond business in the monograph *Good to Great and the Social Sectors.*

An overarching theme across our research findings is the role of discipline in separating the great from the mediocre. True discipline requires the independence of mind to reject pressures to conform in ways incompatible with values, performance standards, and long-term aspirations. The only legitimate form of discipline is self-discipline, having the inner will to do whatever it takes to create a great outcome, no matter how difficult. When you have disciplined people, you don't need hierarchy. When you have disciplined thought, you don't need bureaucracy. When you have disciplined action, you don't need excessive controls. When you combine a culture of discipline with an ethic of entrepreneurship, you create a powerful mixture that correlates with great performance.

To build an enduring great organization—whether in the business or social sectors—you need disciplined people who engage in disciplined thought and take disciplined action to produce superior results and make a distinctive impact in the world. Then you need the discipline to sustain momentum over a long period of time and to lay the foundations for lasting endurance. This forms the backbone of the framework, laid out as four basic stages:

Stage 1: Disciplined People
Stage 2: Disciplined Thought
Stage 3: Disciplined Action
Stage 4: Building to Last

Each of the four stages consists of two or three fundamental principles. The flywheel principle falls at a central point in the framework, right at the pivot point from disciplined thought into disciplined action. I've provided a brief description of the principles below.

STAGE 1: DISCIPLINED PEOPLE

LEVEL 5 LEADERSHIP

Level 5 leaders display a powerful mixture of personal humility and indomitable will. They're incredibly ambitious, but their ambition is first and foremost for the cause, for the organization and its purpose, not for themselves. While Level 5 leaders can come in many personality packages, they're often self-effacing, quiet, reserved, and even shy. Every good-to-great transition in our research began with a Level 5 leader who motivated people more with inspired standards than inspiring personality. This concept is first developed in the book *Good to Great* and further refined in the monograph *Good to Great and the Social Sectors.*

FIRST WHO, THEN WHAT—GET THE RIGHT PEOPLE ON THE BUS

Those who lead organizations from good to great first get the right people on the bus (and the wrong people off the bus) and then figure out where to drive the bus. They always think *first* about "who" and *then* about "what." When you're facing chaos and uncertainty, and you cannot possibly predict what's coming around the corner, your best "strategy" is to have a busload of people who can adapt and perform brilliantly no matter what comes next. Great vision without great people is irrelevant. This concept is first developed in the book *Good to Great* and further refined in the monograph *Good to Great and the Social Sectors.*

STAGE 2: DISCIPLINED THOUGHT

GENIUS OF THE AND

Builders of greatness reject the "Tyranny of the OR" and embrace the "Genius of the AND." They embrace both extremes across a number of dimensions at the same time. For example, creativity AND discipline, freedom AND responsibility, confront the brutal facts AND never lose faith, empirical validation AND decisive action, bounded risk AND big bets, productive paranoia AND a bold vision, purpose AND profit, continuity AND change, short term AND long term. This concept is first introduced in the book *Built to Last* and further developed in the book *Good to Great.*

CONFRONT THE BRUTAL FACTS—THE STOCKDALE PARADOX

Productive change begins when you have the discipline to confront the brutal facts. The best mind frame to have for leading from good to great is represented in the Stockdale Paradox: Retain absolute faith that you can and will prevail in the end, regardless of the difficulties, *and at the same time*, exercise the discipline to confront the most brutal facts of your current reality, whatever they might be. This concept is fully developed in the book *Good to Great.*

THE HEDGEHOG CONCEPT

The Hedgehog Concept is a simple, crystalline concept that flows from deep understanding about the intersection of the following three circles: (1) what you're deeply passionate about, (2) what you can be the best in the world at, and (3) what drives your economic or resource engine. When a leadership team becomes fanatically disciplined in making decisions consistent with the three circles, they begin to generate momentum toward a good-to-great inflection. This includes not only the discipline of what *to* do, but equally, what *not* to do and what to *stop* doing. This concept is first developed in the book *Good to Great* and further refined in the monograph *Good to Great and the Social Sectors.*

STAGE 3: DISCIPLINED ACTION

THE FLYWHEEL

No matter how dramatic the end result, building a great enterprise never happens in one fell swoop. There's no single defining action, no grand program, no one killer innovation, no solitary lucky break, no miracle moment. Rather, the process resembles relentlessly pushing a giant, heavy flywheel, turn upon turn, building momentum until a point of breakthrough, and beyond. To maximize the flywheel effect, you need to understand how your *specific* flywheel turns. The flywheel effect is first developed in the book *Good to Great*, and its application is fully developed in this monograph.

20 MILE MARCH

Companies that thrive in a turbulent world self-impose rigorous performance marks to hit with relentless consistency—like walking across a gigantic continent by marching at least twenty miles a day, every day, regardless of conditions. The march imposes order amidst disorder, discipline amidst chaos, and consistency amidst uncertainty. For most organizations, a one-year 20 Mile March cycle works well, although it could be shorter or longer. But whatever the cycle, the 20 Mile March requires both short-term focus (you have to hit the march *this* cycle) *and* long-term building (you have to hit the march *every* subsequent cycle for years to decades). As such, it's a rarified form of disciplined action that correlates strongly with achieving breakthrough performance and sustaining flywheel momentum. This concept is fully developed in the book *Great by Choice.*

FIRE BULLETS, THEN CANNONBALLS

The ability to *scale* innovation—to turn small, proven ideas (bullets) into huge successes (cannonballs)—can provide big bursts of momentum. First, you fire bullets (low-cost, low-risk, low-distraction experiments) to figure out what will work—calibrating your line of sight by taking small shots. Then, once you have empirical validation, you fire a cannonball (concentrating resources into a big bet) on the calibrated line of sight. Calibrated cannonballs correlate with outsized results; uncalibrated cannonballs correlate with disaster. Firing bullets, then cannonballs, is a primary mechanism for expanding the scope of an organization's Hedgehog Concept and extending its flywheel into entirely new arenas. This concept is fully developed in the book *Great by Choice.*

STAGE 4: BUILDING TO LAST

PRODUCTIVE PARANOIA

The only mistakes you can learn from are the ones you survive. Leaders who navigate turbulence and stave off decline assume that conditions can unexpectedly change, violently and fast. They obsessively ask,

"What if? What if? What if?" By preparing ahead of time, building reserves, preserving a margin of safety, bounding risk, and honing their discipline in good times and bad, they handle disruptions from a position of strength and flexibility. Productive paranoia helps inoculate organizations from falling into the five stages of decline that can derail the flywheel and destroy an organization. Those stages are (1) Hubris Born of Success, (2) Undisciplined Pursuit of More, (3) Denial of Risk and Peril, (4) Grasping for Salvation, and (5) Capitulation to Irrelevance or Death. Productive paranoia is fully developed in the book *Great by Choice,* and the five stages of decline are fully developed in the book *How the Mighty Fall.*

CLOCK BUILDING, NOT TIME TELLING

Leading as a charismatic visionary—a "genius with a thousand helpers" upon whom everything depends—is time telling. Shaping a culture that can thrive far beyond any single leader is clock building. Searching for a single great idea upon which to build success is time telling. Building an organization that can generate many great ideas is clock building. Leaders who build enduring great companies tend to be clock builders, not time tellers. For true clock builders, success comes when the organization proves its greatness not just during one leader's tenure but also when the *next* generation of leadership further increases flywheel momentum. This concept is fully developed in the book *Built to Last.*

PRESERVE THE CORE/STIMULATE PROGRESS

Enduring great organizations embody a dynamic duality. On the one hand, they have a set of timeless core values and core purpose (reason for being) that remain constant over time. On the other hand, they have a relentless drive for progress—change, improvement, innovation, and renewal. Great organizations understand the difference between their core values and purpose (which almost never change), and operating strategies and cultural practices (which endlessly adapt to a changing world). The drive for progress often manifests in BHAGs (Big Hairy

Audacious Goals) that stimulate the organization to develop entirely new capabilities. Many of the best BHAGs come about as a natural extension of the flywheel effect, when leaders imagine how far the flywheel can go and then commit to achieving what they imagine. This concept is first developed in the book *Built to Last* and further developed in the book *Good to Great.*

10X MULTIPLIER

RETURN ON LUCK

Finally, there's a principle that amplifies all the other principles in the framework, the principle of return on luck. Our research showed that the great companies were not generally luckier than the comparisons—they didn't get more good luck, less bad luck, bigger spikes of luck, or better timing of luck. Instead, they got a higher *return* on luck, making more of their luck than others. The critical question is not, will you get luck? But what will you *do* with the luck that you get? If you get a high return on a luck event, it can add a big boost of momentum to the flywheel. Conversely, if you are ill-prepared to absorb a bad-luck event, it can stall or imperil the flywheel. This concept is fully developed in the book *Great by Choice.*

THE OUTPUTS OF GREATNESS

The above principles are the *inputs* to building a great organization. You can think of them almost as a "map" to follow for building a great company or social-sector enterprise. But what are the *outputs* that define a great organization? Not how you get there, but what *is* a great organization—what are the criteria of greatness? There are three tests: *superior results*, *distinctive impact*, and *lasting endurance.*

SUPERIOR RESULTS

In business, performance is defined by financial results—return on invested capital—and achievement of corporate purpose. In the social sectors, performance is defined by results and efficiency in delivering on the social mission. But whether business or social, you must achieve top-flight results. To use an analogy, if you're a sports team, you must win championships; if you don't find a way to win at your chosen game, you cannot be considered truly great.

DISTINCTIVE IMPACT

A truly great enterprise makes such a unique contribution to the communities it touches and does its work with such unadulterated excellence that, if it were to disappear, it would leave a gaping hole that could not be easily filled by any other institution on the planet. If your organization went away, who would miss it, and why? This does not require being big; think of a small but fabulous local restaurant that would be terribly missed if it disappeared. Big does not equal great, and great does not equal big.

LASTING ENDURANCE

A truly great organization prospers over a long period of time, beyond any great idea, market opportunity, technology cycle, or well-funded program. When clobbered by setbacks, it finds a way to bounce back stronger than before. A great enterprise transcends dependence on any single extraordinary leader; if your organization cannot be great without you, then it is not yet a truly great organization.

Finally, I caution against ever believing that your organization has achieved ultimate greatness. Good to great is never done. No matter how far we have gone or how much we have achieved, we are merely good relative to what we can do next. Greatness is an inherently dynamic process, not an end point. The moment you think of yourself as great, your slide toward mediocrity will have already begun.

NOTES

[1]Stephen Ressler, *Understanding the World's Greatest Structures* (Chantilly, VA: The Teaching Company, 2011), Lecture 24.

[2]Brad Stone, *The Everything Store* (New York, NY: Little, Brown and Company, 2013), 6-8, 12, 14, 100-102, 126-128, 188, 262-263, 268.

[3]Erika Fry, "Mutual Fund Giant Vanguard Flexes Its Muscles," *Fortune*, December 8, 2016, http://fortune.com/vanguard-mutual-funds-investment/; "Fast Facts about Vanguard," *The Vanguard Group, Inc*, Accessed in 2017, https://about.vanguard.com/who-we-are/fast-facts/.

[4]Robert N. Noyce, "MOSFET Semiconductor IC Memories," *Electronics World*, October 1970, 46; Gene Bylinsky, "How Intel Won Its Bet on Memory Chips," *Fortune*, November 1973, 142-147, 184; Robert N. Noyce, "Innovation: The Fruit of Success," *Technology Review*, February 1978, 24; "Innovative Intel," *Economist*, June 16, 1979, 94; Michael Annibale, "Intel: The Microprocessor Champ Gambles on Another Leap Forward," *Business Week*, April 14, 1980, 98; Mimi Real and Robert Warren, *A Revolution in Progress...A History of Intel to Date* (Santa Clara, CA: Intel Corporation, 1984), 4; Gordon E. Moore, "Cramming More Components onto Integrated Circuits," *Proceedings of the IEEE*, January 1998, 82-83; Leslie Berlin, *The Man Behind the Microchip* (New York, NY: Oxford University Press, 2005), 160, 170-172; "Moore's Law," *Intel Corporation*, Accessed in 2018, http://www.intel.com/technology/mooreslaw/.

[5]Andrew S. Grove, *Only the Paranoid Survive: How to Exploit the Crisis Points that Challenge Every Company* (New York, NY: Crown Business; 1st Currency Pbk. Ed edition, April 23, 2010), 85-89.

[6]James C. Collins and William C. Lazier, *Managing the Small to Mid-Sized Company* (New York, NY: Richard D. Irwin Publishers, 1995), C47-C74.

[7]James C. Collins and William C. Lazier, *Managing the Small to Mid-Sized Company* (New York, NY: Richard D. Irwin Publishers, 1995), C47-C74.

[8]Jim Collins and Morten T. Hansen, *Great by Choice: Uncertainty, Chaos and Luck—Why Some Thrive Despite Them All* (New York, NY: HarperBusiness, 2011), 76; Brad Stone, *The Everything Store* (New York, NY: Little, Brown and Company, 2013), 34.

[9]Gerard J. Tellis and Peter N. Golder, *Will & Vision* (New York, NY: McGraw-Hill, 2002), 257; Jim Collins, *Good to Great: Why Some Companies Make the Leap and Others Don't* (New York, NY: HarperCollins Publishers Inc., 2001), 149, 152, 158; Jim Collins and Morten T. Hansen, *Great by Choice: Uncertainty, Chaos and Luck—Why Some Thrive Despite Them All* (New York, NY: HarperBusiness, 2011), 70-71, 72-73, 76, 89-90, 168; Jim Collins and Jerry I. Porras, *Built to Last: Successful Habits of Visionary Companies* (New York, NY: HarperBusiness, 1994), 25-26.

[10]Author interview with Deb Gustafson; Karin Chenoweth, "The Homework Conundrum," *The Huffington Post*, March 12, 2014, http://www.huffingtonpost.com/Karin-Chenoweth/the-homework-conundrum_b_4942273.html.

[11]Author interview with Deb Gustafson; Karin Chenoweth, "How it's Being Done: Urgent Lessons from Unexpected Schools – Student Services Symposium," *The Education Trust*, May 17, 2010, 24-26.

[12]Author interview with Deb Gustafson.

[13]"About," *Ojai Music Festival*, Accessed in 2018, https://www.ojaifestival.org/about/; "Milestones," *Ojai Music Festival*, Accessed in 2018, https://www.ojaifestival.org/about/milestones/.

[14]Author interview with Tom Morris.

[15]"Inuksuit, John Luther Adams, and Ojai," *Ojai Music Festival*, Accessed in 2018, https://www.ojaifestival.org/inuksuit-john-luther-adams-and-ojai/.

[16]Author interview with Tom Morris.

[17]Author interview with Dr. Toby Cosgrove; Toby Cosgrove, *The Cleveland Clinic Way: Lessons in Excellence from One of the World's Leading Health Care Organizations* (New York, NY: McGraw-Hill Education, 2013).

[18]Author interview with Dr. Toby Cosgrove; Toby Cosgrove, *The Cleveland Clinic Way: Lessons in Excellence from One of the World's Leading Health Care Organizations* (New York, NY: McGraw-Hill Education, 2013); "Toby Cosgrove, M.D., Announces His Decision to Transition from President, CEO Role," *Cleveland Clinic*, May 1, 2017, https://newsroom.clevelandclinic.org/2017/05/01/toby-cosgrove-m-d-announces-decision-transition-president-ceo-role/; Lydia Coutré, "The Cosgrove Era Comes to a Close," *Cleveland Business*, December 10, 2017, http://www.crainscleveland.com/article/20171210/news/145131/cosgrove-era-comes-close.

[19]Jim Collins and Morten T. Hansen, *Great by Choice: Uncertainty, Chaos and Luck—Why Some Thrive Despite Them All* (New York, NY: HarperBusiness, 2011), Chapter 4.

[20]Jim Collins and Morten T. Hansen, *Great by Choice: Uncertainty, Chaos and Luck—Why Some Thrive Despite Them All* (New York, NY: HarperBusiness, 2011), 91-95.

[21]Amazon, *Fiscal 2015 Annual Letter to Shareholders* (Seattle, WA: Amazon, 2015); Amazon, *Fiscal 2016 Annual Report* (Seattle, WA: Amazon, 2016); Amazon, *Fiscal 2017 Annual Report* (Seattle, WA: Amazon, 2017); Alex Hern, "Amazon Web Services: the secret to the online retailer's future success," *The Guardian*, February 2, 2017, https://www.theguardian.com/technology/2017/feb/02/amazon-web-services-the-secret-to-the-online-retailers-future-success; Robert Hof, "Ten Years Later, Amazon Web Services Defies Skeptics," *Forbes*, March 22, 2016, https://www.forbes.com/sites/roberthof/2016/03/22/ten-years-later-amazon-web-services-defies-skeptics/#244356466c44.

[22]Amazon, *Fiscal 2017 Annual Report* (Seattle, WA: Amazon, 2017); "Global Retail Industry Worth USD 28 Trillion by 2019 - Analysis, Technologies & Forecasts Report 2016-2019 - Research and Markets," *Business Wire*, June 30, 2016, https://www.businesswire.com/news/home/20160630005551/en/Global-Retail-Industry-Worth-USD-28-Trillion.

[23]Jim Collins, *How the Mighty Fall: And Why Some Companies Never Give In* (Boulder, CO: Jim Collins, 2009), 29-36.

[24]Alan Wurtzel, *Good to Great to Gone: The 60 Year Rise and Fall of Circuit City* (New York, NY: Diversion Books, Kindle Edition, 2012), Chapter 8; Circuit City Stores, Inc., *Fiscal 2002 Annual Report* (Richmond, VA: Circuit City Stores, Inc., 2002); Michael Janofsky, "Circuit City Takes a Spin at Used Car Marketing," *The New York Times*, October 25, 1993, http://www.nytimes.com/1993/10/25/business/circuit-city-takes-a-spin-at-used-car-marketing.html; Mike McKesson, "Circuit City at Wheel of New Deal for Used-Car Shoppers: Megastores," *Los Angeles Times*, January 28, 1996, http://articles.latimes.com/1996-01-28/news/mn-29582_1_circuit-city.

[25]Alan Wurtzel, *Good to Great to Gone: The 60 Year Rise and Fall of Circuit City* (New York, NY: Diversion Books, Kindle Edition, 2012), Chapter 8; Circuit City Stores, Inc., *Fiscal 2002 Annual Report* (Richmond, VA: Circuit City Stores, Inc., 2002).

[26]Alan Wurtzel, *Good to Great to Gone: The 60 Year Rise and Fall of Circuit City* (New York, NY: Diversion Books, Kindle Edition, 2012), Loc 3542 of 5094.

[27]Alan Wurtzel, *Good to Great to Gone: The 60 Year Rise and Fall of Circuit City* (New York, NY: Diversion Books, Kindle Edition, 2012), Chapter 8 and Chapter 10; Jesse Romero, "The Rise and Fall of Circuit City," *Federal Reserve Bank of Richmond*, 2013; Jim Collins, *How the Mighty Fall: And Why Some Companies Never Give In* (Boulder, CO: Jim Collins, 2009), 29-36.

ABOUT THE AUTHOR

Jim Collins is a student and teacher of what makes great companies tick, and a Socratic advisor to leaders in the business and social sectors. Having invested more than a quarter-century in rigorous research, he has authored or coauthored six books that have sold in total more than 10 million copies worldwide. They include *Good to Great*, *Built to Last*, *How the Mighty Fall*, and *Great by Choice*.

Driven by a relentless curiosity, Jim began his research and teaching career on the faculty at the Stanford Graduate School of Business, where he received the Distinguished Teaching Award in 1992. In 1995, he founded a management laboratory in Boulder, Colorado.

In addition to his work in the business sector, Jim has a passion for learning and teaching in the social sectors, including education, healthcare, government, faith-based organizations, social ventures, and cause-driven nonprofits.

In 2012 and 2013, he had the honor to serve a two-year appointment as the Class of 1951 Chair for the Study of Leadership at the United States Military Academy at West Point. In 2017, Forbes selected Jim as one of the 100 Greatest Living Business Minds.

Jim has been an avid rock climber for more than forty years and has completed single-day ascents of El Capitan and Half Dome in Yosemite Valley.

Learn more about Jim and his concepts at his website, where you'll find articles, videos, and useful tools. jimcollins.com

HARPER
BUSINESS

Available from HarperAudio and HarperCollins e-books

Discover great authors, exclusive offers, and more at hc.com

Cover and text design by Elements Design Group